The Building of a Wooden Ship

The Building of

TEXT BY DANA A. STORY

PHOTOGRAPHS BY JOHN M. CLAYTON

a Wooden Ship

"Sawn Frames and Trunnel Fastened"

INTRODUCTION BY PHILIP C. F. SMITH

BARRE PUBLISHERS Barre, Massachusetts 1971

Library of Congress Catalog Card Number 72-163015

Standard Book Number 8271-7115-3

Printed in the United States of America

Introduction

ANYBODY who has ever seen one knows that the launching of a wooden vessel is a dramatic event. A jovial crowd in a festive mood lines the shore. Anticipation heightens as the shipyard hands are seen to re-examine the ground ways and here or there, as necessary, slap another dob of grease over the "holidays." At last, teams of men start splitting out the blocks under the keel or saw away like madmen to the marks at the timber-ends that restrain the sliding ways. An unnatural silence falls in a wave across the spectators. The great hull begins to tremble ever so slightly. The sponsor swings her beribboned bottle of champagne and sometimes discovers to her surprise that a bottle is tougher than she had supposed and that her first swat had been too ladylike, so she throws her weight into the stroke and gets it this time. There is a crack of timber, the creaking groan of wood against wood as the vessel creeps, the splitters or sawyers bound aside, and a shrill chorus of voices screeches in unison, "There she goes!" The hull gathers momentum. It dashes down the ways in a welter of greasy smoke. The drag thunders along behind. The new creature dips, bobs, and is waterborne. Motor cars and small boats hoot their horns in jubilation, and there is a good deal of frenzied cap-waving from the assembled audience. Any cameramen on hand by this time are nearly beside themselves.

Dramatic events, especially predictable ones, make good pictures, and wherever good pictures are possible there are sure to be photographers. From the moment photographic equipment became sufficiently inexpensive and portable to appeal to the amateur, the shipyards of Essex, Massachusetts, always had their share of enthusiasts.

By and large the results fell into two categories—the spray-soaked, soul-stirring launching shot and the tranquil Sunday afternoon scenes of unpopulated shipyards in temporary hibernation. Not very many people seemed to think or care about what happened during the week or how the men went about doing their work. I suppose most everyone took for granted what they accomplished and thought Essex would always be building her ships.

As a boy I was driven past the shipyards on numerous occasions—Essex was usually on the guided-tour circuit my father loved to give to out-of-town friends. I don't know how they felt about those fantastic skeleton shapes that cast such ragged shadows, but they certainly made an impression on me. To this day I cannot go to Essex without half expecting to turn a corner and see them again. I know they won't be there, yet I always hope they might be.

One of the first pioneer cameramen on the Essex scene was Edwin J. Story, to whom we owe much for his splendid glass plate negatives which recorded the yards during the early decades of this century. Even he, however, only infrequently set up his camera in front of working men. He couldn't. He was too busy himself in the mold loft. To the best of my knowledge he never zoomed in on a vessel for the purpose of detailing the construction. Among those who did was John Clayton, who is largely responsible for this book's pictorial content.

Mr. Clayton's skill with the camera, his love of the yards, and his methodical documentation of practically everything he saw there over a period of years resulted in a collection so complete that it is possible to reconstruct many all-but-forgotten details of an industry which sustained the town from the mid-seventeenth century until the years immediately following World War II. No plans in a museum could do the same, because no plans, no matter how meticulously drafted, can anticipate the multitude of thorny little problems that arise on the actual job, nor can they attempt to show how a particular shipbuilding community (let alone individual builders) traditionally chose to fashion their vessels. Now that the industry has vanished from Essex, virtually without trace, we are left only with the memory of it and a host of pictures. We would be happy to have the photographs in any event, but, if we can't have the real thing, how lucky we are to have had such men as John Clayton!

How fortunate we are, too, to have benefited now and in the past from Dana Story's beguiling pen. The youngest son of the famous Arthur D. Story—builder of 425 vessels including such immortal racing fishermen as *Elsie, Henry Ford, Columbia,* and *Gertrude L. Thebaud*—Dana Story has himself built wooden ships and knows them well. His two previous books, *Frame-Up!* and *Hail Columbia* published by Barre, have perpetuated the memory of Essex men and their ships with a racy flavor hard to beat. The cast of characters in this book is much the same, but this time we can follow along with them as they work. Look at them well. Note the presence of so many remarkable septuagenarians and octogenarians pulling their weight with the rest of the boys (and no doubt besting the youngsters every time).

As you look at these pictures and read their descriptions, let yourself melt into them. Let yourself walk on generations of accumulated oak chips and pine shavings. They feel springy underfoot like soft, deep beds of moss, and they smell sweet and pungent—almost heady. As Jack Conrad hauls back with his beetle to drive a trunnel home, feel the heft of it. Listen to the sound. It is not the same sound as an adze-man squaring off a timber on the ground; there is a satisfying, deep-throated resonance to it. Can you wonder why caulkers were prone to premature deafness when you hear the sharp, hollow ringing of Bill Boutchie's mallet striking his iron? In the background a planing machine pierces the air with a long scream. Wispy trails of vapor escape from the steam box. Someone drops a crowbar on a deck. Voices emerge from unseen corners. Every once in a while all the sounds cease for a moment. If you listen carefully you might then hear a camera click.

Philip Chadwick Foster Smith
Curator of Maritime History

Peabody Museum of Salem
Salem, Massachusetts

Foreword

SHIPBUILDING was an industry which became established in what is now Essex, Massachusetts, back in the middle 1650s. The topography of the region, together with its soil characteristics, made it relatively inhospitable to agriculture on any large scale, so that, somewhat of necessity, the early settlers turned to the sea to augment their food supplies. Being blessed, however, with magnificent stands of virgin timber, of both hard and soft woods, they used these forests for lumber with which to build the boats employed in harvesting the sea. Thus was born a shipbuilding industry which grew to considerable prominence over the years. It reached a point where in the 1850s there were as many as sixteen shipyards in the little town, turning out over sixty vessels a year. Over the years it has been estimated that something over 4,000 ships have come from the yards of Essex.

The building of a wooden ship, whether large or small, was a hand process, and with minor exceptions has remained so to the present day. Its nature does not easily lend itself to mechanization, automation, or prefabrication. As a hand process it was (and is) extremely interesting to watch, a fact attested by the hordes of people who, as automobiles and better roads became commonplace, stopped by the Essex yards to observe what was taking place.

One of those who came to Essex and was fascinated by the sights of an old-time wooden shipyard was John Clayton. Being an able photographer and also one whose business was directly concerned with photography, he began to record in minute photographic detail the processes and operations by which a wooden ship was built. The result was a collection of many thousands of priceless and historic negatives which, as it turned out, also recorded the final years of the once great Essex shipbuilding industry.

From these many negatives we have attempted to make a small selection of pictures which illustrate in a very general way the growing process of a ship, and, perhaps more especially, to depict life in the shipyard: to show the men, what they were doing, and why they were doing it. Most of the photographs span a period of approximately nine years, beginning in 1938, and show work in process on perhaps as many as ten different boats in six different yards. Actually, to say "six different yards" is somewhat misleading, since only three sites are involved, but one of them is shown at various times under four different Story managements: A. D. Story, Jacob Story, John Prince Story, and D. A. Story. John Prince Story, who hired this yard in 1941 to build the *Theresa M. Boudreau,* moved the next year to what had been an old Burnham yard and continued his business in company with his grandson Jonathan Story from that site. The third site, the James yard, had been under the management of the James family for generations.

Because they seemed particularly appropriate to our purposes, many of the photographs illustrate work upon a small 55-foot schooner yacht, the *Blackfish,* built by Jacob Story in the spring and summer of 1938. Another large group shows work upon the dragger *Ronald and Mary Jane,* built by Lyman James in the spring and summer of 1941. The rest are divided among several vessels being built by John P. Story, Jacob Story, Dana Story, Jonathan Story, and Lyman James over the years 1938 to 1947. Two or three go back to the early years of the century in the old A. D. Story yard. In any case, whether for large dragger or small schooner, whether 1900 or 1947, the processes illustrated are essentially the same.

Each of the boats shown had frames sawn to shape (as distinguished from bent to shape) and with the exception of the *Blackfish,* the planking on all was fastened by "trunnels" or wooden pins. The word is a corruption of the word "treenail." To a boatman the descriptive phrase "sawn frames and trunnel fastened" implied but one thing—a good, rugged boat. The reputation of Essex shipbuilders was built upon one thing—good, rugged boats.

Before beginning the series of pictures it might be well to call one final point to the reader's attention. Understandably, as the last years of the industry approached, the number of old-time shipwrights competent to perform nearly any operation dwindled noticeably, with the result that their services, even though many were men of advanced years, were earnestly sought by all yards. Consequently, there are a number of photographs which show the same men at work in several yards at various times.

Let us now leave further explanation to the pictures themselves and go back to old Essex and the days when they were building vessels there.

THE story that we hope to tell in this book of pictures is neatly contained in a photograph of the John Prince Story yard taken in 1946.

To the right is a brand new keel. At the left is a large fishing dragger nearly finished and ready for the water. The processes by which the transition from one to the other is made will be the object of our pictorial narrative.

IN common with all creation, the building of a ship must start somewhere. For generations—indeed, for centuries—the shape of a boat has been originally conceived in the form of a model. Having achieved in the model the contours which he feels are the ones he desires, the boatbuilder proceeds to transpose the shape of the scale model to the full size of his boat.

Viewed in its more practical aspects, the two sides of a ship must (at least theoretically) be exactly alike. Therefore, one need only make a model of half the boat, saving thereby a great deal of needless labor and creating in so doing the commonly designated "half-model."

Here we see David Mulcahy running his fingers over the nearly completed half-model of a fishing "dragger" about to be built in Lyman James' Essex shipyard. The model, of clear white pine, is made in layers or "lifts" fastened together with dowels so that it may be taken apart. By tracing the lifts, the model thus becomes delineated upon paper and a set of "lines" evolved which can then be enlarged to the full size of the ship.

WITH the completion of a "linesdrawing" taken from a half-model, it now becomes necessary to enlarge this small scale drawing to the full size of the contemplated vessel. This is done on the floor of a "mold loft," where not only the shape of the hull but all of the principal structural parts are drawn full size. Now from the full-size drawings, templates or "molds" of these pieces must be made. The man who in Essex made this his profession for many years and who made the molds for an uncounted number of vessels was Archer B. Poland, whose loft was over on School Street in South Essex.

Here we see Mr. Poland kneeling on the paper-covered floor as he draws the sections of the new vessel. From his drawings he will make the molds out of thin pine boards, some of which we see on the floor by his band saw. Mr. Poland designed many of the boats he lofted and we can see a half-model and some of the implements of the marine draftsman on the bench along the back wall. We might also note the astonishing variety of miscellaneous objects with which Mr. Poland's loft was garnished.

HERE is a picture of "Doc" Stanwood and David Mulcahy using the device (sometimes called a "trammel") by means of which the lines on the floor were transferred to pine boards of the mold. The body of the thing was fixed in place with awls punched into the floor. Then the adjustable arms were moved to hold a slender batten along the desired line. The batten could be raised a little and the template stock slid under it while the line was traced on. Quite simple, really.

WITH the model decided upon and made, the lines drawn and molds ready, the time has come to begin the actual construction of the vessel. As might perhaps be imagined, the place to begin is with the backbone or, more properly, the keel of the boat.

Face to face in what appears to be something of a confrontation is John Prince Story, the master builder and foreman, and Stephen Price, a veteran shipwright of long experience. Both men are well into their eighties. Between them on the ground is the first piece of a new keel. Maybe they're discussing whether or not it will be all right to leave those two little edges of bark (or "wane") on the upper corner.

Mr. Story was much better known to all as "John Prince," which in Essex served to distinguish him from the multitude of other Storys who in years past inhabited the place. He was not at all as belligerent as the outthrust chin seems to imply.

SIMPLY because the nature of its structure makes it so, a keel is more easily assembled upside down. When finished it is rolled over and carefully placed on the blocks upon which the growing vessel will sit until it is launched.

Here, several pieces of keel are fastened together and another is waiting to be cut and fitted. Showing is one of the tapered cuts or "scarphs" by which the pieces were joined.

NOT all keels, of course, were alike and not all were assembled in quite the same way. Many of the pictures in our series will depict work on a lovely little schooner yacht. The first of these is this one, which shows Leander Doucette using his adze to trim the large central oaken member to which will be fastened the heavy iron casting forming the outside ballast of the boat. This piece has been roughed into shape on the big band saw and now Leander is bringing it into its finishd form, a degree of finish somewhat akin to a grand piano. Leander could, and usually did, split the pencil mark which here clearly shows on the side of the piece, and his finished surface was like glass. Old-time axe and adzemen would blanch in horror at the sight of our popular modern "hand-hewn" beams. Any man who could do no better than to leave a surface as rough as these contemporary pieces would turn in his tools and take to raising chickens.

HERE'S how the keel of a big dragger looks when it's finished and finally in place. The top layer is the one actually called the keel and is 12″ x 14″. Below it is the so-called "false keel" which is 12″ x 10″. On the bottom is the "shoe" and that's 12″ x 3″. Here Clayt McKay and Fred Manthorne are setting some side or "dog" shores to hold the keel in perfect alignment. Spiked along the top are some cleats for the platform on which the frames of the vessel will be assembled.

WITH the keel in place, a vessel dramatically begins to take shape as the heavy frames are put together and raised into place. Here is a splendid photo of work in progress on the "framing stage" which illustrates clearly how the several parts called "futtocks," each with its own name, are fitted together and molded into the whole. The pieces were fastened together either with iron drift pins or wooden trunnels. As interesting as the frame itself is the assortment of tools and oddments of the framing process lying about the platform. No doubt those big C-clamps, or "screws" as they were commonly called, were performing the same function fifty years earlier. We might add that "Old Tom" Boutchie, the left-hand figure in the picture, had also performed this same function nearly as long ago and was himself well over eighty at the moment here recorded.

WITH the frame completely assembled and fastened (in this case with trunnels because we see the heads sticking out) a "cross pawl" is spiked across the tops of the frame to keep it from spreading, and it's ready to stand up. "Frame-up!" holler the fellows on the framing stage, and the rest of the gang drop whatever they're doing and come over to help. Stooping over together they grab the thing and stand it up at its place on top of the keel. It's awfully heavy too, since, like all structural members, it's made of oak and each of the two layers composing it is, in this case, six inches thick.

ONCE a frame is standing up, some of the fellows must balance it in place while the rest jiggle or jostle or pound it into just the right spot and then attach a few temporary pieces to hold it there. Having completed this, the rest of the gang depart and the framers start their little cycle all over again.

THE skeleton of a ship is composed of many parts. While some of the gang are assembling frames, others are getting out the timbers which will form the ends of the vessel. The keel as it curves upward to form the bow becomes the "stem" as is itself made up of several heavily timbered members. Since Mother Nature does not produce all perfect trees, so all of the heavy timbers of a vessel are not without a blemish or two whether they be keel, frames, stem, or whatever. If an otherwise suitable piece contains a small black knot or a nest hole of a brown-tail moth, the offending spot is cut out and a "graving piece" put in. Here we see "Old John" Hubbard fitting a graving piece into a big stem timber. It would appear that another will be required in the face of the stem under his right hand.

SOME idea of what we mean when we say a stem is composed of several heavily timbered members can be gained when we see its bulk soaring over the heads of Cliff McKay and Charlie LeBlanc as they work at cutting in the "rabbet." The rabbet is not a four-legged creature but the V-shaped slot into which the vessel's planking will fit as it joins onto the stem (or onto the keel or stern structures as well). Obviously this is the stem of a fairly large vessel, and, while others may not have been as big, all of them were bulky and heavy and painstaking to erect.

WE have been concerned thus far with procedures taking place on the growing vessel itself. Manifestly other operations have been going on behind the scenes in support of those we have illustrated. For example, somebody had to get out all those pieces making up the frames, since the only thing the gang on the framing stage did was put them together. It was in preparing these timbers that the molds made by Archie Poland were used. The man who "molds" timber takes his templates and selects appropriate pieces from his stock pile. He must remember the sequence of pieces and number them correctly; he must decide where to make joints; he must indicate the inside edge and angle of cut and he must use his available stock to greatest advantage. With this done, the raw stock goes to the band saw, where the sawyer cuts it out.

Here is the sawyer, Sammy Gray. (There were people who knew Sammy Gray for years without ever realizing that his real name was not Samuel but George P. Gray.) As it turns out, the piece Sammy happens to be sawing in the picture is not a section of frame but rather appears to be part of a stern or stem.

WE have discussed how frames are put together and raised. The type we have seen and which is put up in a single unit is known as a "square frame." These formed the central body of the ship. As the shape of the hull becomes finer and somewhat more complex at the ends of the vessel, the frames are erected in halves, one side at a time. These frames, whether at the forward or after ends of the vessel, are called "cants." This is because in the early years of wooden shipbuilding they were "canted" or set at an angle to the centerline of the keel.

Here the central or square frames of the little schooner are up and it's time to erect the forward half-frames or cants. Banging away with his adze as he trims up the stem in preparation for raising the forward cants is John Mulcahy, another master shipwright of the old tradition. Half hidden by the staging, somebody else below him is busily cutting the rabbet. That long piece which connects the top of the square frames with the stem is called a "harpin" and it supports and aligns the top of the cants as they are raised.

THIS is what it takes to hoist the cants into place—"brute strength and ignorance," as we used to say. It isn't a great deal different from almost any other process in the old-time shipyard. With perhaps one or two exceptions, cranes or derricks in the usual sense of the word were never used in the Essex yards, and heavy members were largely manhandled into place or hoisted with blocks and falls as these men are doing under the watchful eye of foreman John Prince. These are not young men doing this, either.

HERE we see it again, only now we're at the after end of the vessel and the men are starting to erect the stern structure. This, as might be imagined, is much more involved and the structure considerably more complex than the bow. The piece being lifted here is the stern post, a member of considerable bulk and weight. The hole through which that old hand-wrought chain passes will later receive the vessel's propeller shaft. Interestingly, the man pulling on the falls is our friend Steve Price whom we saw earlier working on a keel, only here he's nearly a year older. Steve was one of those rare individuals who never seemed to grow old; with his new cap, jumper, and overalls on, he looks positively dapper. On this occasion, Steve is working for Lyman James.

THIS is a splendid illustration of the emerging after-structure of the dragger. The stern post which we saw being lifted into place is up now, fastened and shored. Behind it stands the even bigger rudder post, still supported by the two-legged derrick or "shears" that was used to lift it there. Such a device was used in Essex until the end to lift the tall and/or heaviest members into place. For a piece too heavy to raise with just the men pulling on the falls, a crude wooden capstan called a "crab" was employed, wherein the vertical drum was turned by a long wooden timber projecting on either side like spokes and the men walking it around and around. If even that was not enough, the shipyard horse was hooked on. No doubt the stones of the pyramids were placed in much the same way.

Over to the left, preparing the timbers for the transom, are Willy Ross, David Mulcahy, and Steve Price.

A STERN post of a somewhat different character is being placed in this photograph. This one happens to be for a smallish double-ended yacht. Note, however, that the means of raising it is the same, although an insurance inspector might wince at the sight of those shear legs reinforced by pieces held on with a few C-clamps. Of particular interest perhaps is the picture of that big stern mold which, when set on the keel, will give the exact position of the stern post and indicate the location of the rest of the stern structure. For a "double-ender" this stern post roughly compares with the stem at the forward end of the vessel.

Standing in the shadow, the stump of a cigar clamped in his teeth, is the ever-present John Prince keeping track of the proceedings. Behind him, across the creek, stand other partially completed draggers in yet another Story yard.

DAVID MULCAHY, Steve Price, and "Doc" Stanwood are getting along well with the building of that big stern. The horn timber connecting stern post with rudder post is in, and running by on either side are the cheek pieces containing the mortises for the tenons on the heel of each of the after cants. Since the vessel here being worked on is a modified schooner type, she will have a transom stern, and we see the first two timbers of this, the "tailfeathers," already in place.

The men are using the transom mold to erect a sort of jig to hold the transom timbers in position. We see, also, that a gin pole has been set, and the falls are ready to begin raising the various pieces. When the time comes, the rudder stock will pass up through the hole in the rudder post and come out on deck. Sharp eyes will detect here that the after half of the rudder post is made of white pine whereas the post itself is of oak. By the way, not the least of the jobs these men have done is the erection of that high and heavy platform stage from which they can safely work.

THE completion of a vessel's framing presents an interesting pattern of light and symmetry when viewed from inside. It gives us also a look at details of stern structure that cannot be seen from the outside. Those big timbers lying against the keelson are engine beds placed there, with Yankee reason and common sense, when the vessel was framed in order to avoid the necessity of hoisting them up over the side and into the hold after the vessel was complete. We see also that the first couple of streaks of inside planking or ceiling have been put in and note that some trunnels which hold on the outer plank are sticking through.

ONCE the stem structure of a vessel is well along, preparations can be made to start the planking. First of all, this involves cutting the rabbet for the full length of the keel. To repeat, the rabbet is, roughly, a V-shaped notch cut to receive the edge of the first strip or "streak" of plank. This first streak is called the "garboard." Next the "dubbing" of the frames must begin. By dubbing we mean the trimming of each frame with the adze so that the planks will lie firmly and fairly against them.

It appears from this picture that Dennis Martin is doing a little of each.

HERE'S John Prince again. John Prince Story operated his own business on two occasions in his lifetime, the first as a young man in the middle 1890s, following in the footsteps of his father, Joseph. He then worked in several capacities, mostly as a foreman in various Essex yards and for a while in the Bishop yard in Gloucester. In 1941, when in his early eighties, he started his own business again, first building the big Gloucester dragger *Theresa Boudreau*. In this picture we see him "lining plank" on the *Boudreau*. We see also that the *Boudreau* has double sawn 6″ frames (meaning that each frame is made up to two layers of 6″ timbers sawn to shape), and that she is being planked with 3″ oak, trunnel fastened. The experienced observer will note from looking at the timbers that Old John wasted nothing when he molded them.

"Lining plank" means to lay out the strips or "streaks" (sometimes "strakes") of plank which form the outer skin of the ship, and to do it in such a way that they fit fairly and in proper relationship to those adjacent and to those already on. Without going into the details of the process, suffice it to say that what appears to be a simple process actually takes a good deal of experience and acumen. Be assured that John Prince had plenty of both.

Note further that the trunnels fastening the plank have already been sawed off and wedged to make them secure. One can tell because the photo shows the wedges sticking out.

NO SMOKING

THE "mill" was the term we used in Essex to describe the building which housed the big band saw. It also contained the lathe for turning trunnels, an ancient surface planer, and a wonderful big old wet stone. It was an exciting place, especially when the machinery was running, and the noise of the whirling pulleys, flapping belts, and whining band saw were friendly sounds. One somehow didn't even mind the scream of the planer. Perhaps the nicest thing about it was the wonderful smell of fresh chips and sawdust. There seems to be a reassurance in making things out of good wood. In all there was a kind of friendliness to the place, as we can imagine from seeing Bob D'Entremont, John Mulcahy, and Steve Price as they contemplate a plank before sawing it.

Off to the left, Phil Terrio (his name had originally been Theriault) busies himself in smoothing up a stanchion.

LINING out a plank is the first step in its creation; sawing it out is the second. After sawing, it is lugged over to where the "beveller" makes it ready to put onto the vessel. By beveller we mean the man who smooths with broad axe and plane the edges of a plank so that it will lie firmly and fairly against the one below or above it. Bear in mind that the finished edges generally have a constantly varying angle (or bevel) throughout their length.

Here we see the crowd lugging the plank from the mill to wherever it is the beveller is working. By the way, this is not the same plank nor is it even the same point of time we saw in the previous picture.

HERE is a beveller at work. The man doing the bevelling is Leander Doucette, whom you may remember seeing earlier trimming a keel piece. This particular plank is of long leaf yellow pine, commonly called "hard pine." The numbers you see along the upper edge of the plank designate the numbers of the frames against which this particular plank will lie.

Leander Doucette was a remarkable man. As was the case with so many of the shipwrights of Essex, he could turn his hand to just about any phase of the ship's construction and do a good job. Also, in common with many of our Essex shipyard men, he had begun his career as a dory fisherman out of Nova Scotia, and had come to Essex to settle and find thereby a more steady and rewarding employment.

ALTHOUGH we touched upon it briefly earlier, we want to elaborate a little more on what Phil Terrio is doing here. Before any plank is hung onto the vessel, a small flat place has to be trimmed or "dubbed" with an adze from the curvature of each frame. This, of course, is to enable the plank to lie flat against the frame. These frames on the little schooner are small and the dubbing is relatively easy, but imagine the chopping that took place in building the big vessels—a 140-footer, maybe, whose frames may have been fourteen or sixteen inches thick. It took arms of iron.

NO two ways about it, "hanging" a plank on a vessel is a hard and laborious job, especially when it's up in the "tuck" under the overhang of the stern. As you see, this is the area where the overhanging stern comes down to meet the stern post. The shape of the ship here changes almost immediately from vertical to nearly horizontal. If the plank is of 3″ oak (as this one is) and over twenty feet long, it takes a little grunting. These men have raised it on their shoulders to approximately where it belongs, at which point it is secured by heavy iron C-clamps, finally to be wedged snugly against the one below it. Just think what it must have been like in the days when they built the really big wooden ships. The planking in those vessels was four and five inches thick and sometimes fifty or sixty feet long.

AFTER all, it's the planking or "skin" that makes a ship a ship, and it's a fussy job to be sure that each plank lies properly and snugly against the one beneath it. You can't depend on just oakum and cotton to keep you afloat. It's hard, too, when a plank has to be bent edgeways (called "set"), flatways (called "bend"), and also twisted. In such a situation a good cooking in the steam box is necessary. Obviously, then, more than a little skill and muscle is required to hang a heavy plank on a big vessel.

The fastener's job is important too. These men follow behind the planking gang and fasten the plank where it belongs. In this picture we see Bob D'Entremont on the right boring the 1⅛″ diameter holes for the trunnels, four holes per frame per plank. Bigger vessels used 1¼″ or 1⅜″ trunnels. Behind him is Jack Conrad, the trunnel driver. In his hand is the wooden mallet or "beetle" used for driving and just beyond him is a little pile of freshly turned trunnels. The driver turns his own trunnels and it is important that they be just right, neither too loose not too tight. If too tight, they break in driving; if loose, they aren't holding anything. It should take at least eight or nine good blows to drive a trunnel. After driving, the projecting head is sawed off and the trunnel split and wedged to make it secure. Trunnels were customarily made of locust, a hard and durable wood whose rate of expansion when wet was greater than that of the oak it was holding.

In the background John Hubbard is dubbing for the next streak.

HERE is an interesting and rather picturesque view of the little schooner with framing complete and planks about half on. We note that the batten determining the final sheer line is nailed to the timber heads. The man lining plank needs this in laying out the rest of his streaks. By the look of the buds on the towering elm tree it must be about mid-May. By the look of the trash on the ground it's time to do a little picking up.

FOUR pictures ago we saw Phil Terrio dubbing as planking progressed on the little schooner. Here he is again, and now he's getting out the stanchions, those timbers which project above the deck and to which the bulwarks and rails are attached. Stanchions are generally put in separately after the vessel is all planked. Note the well-worn oil stone and two smoothing planes. We see also the traditional carpenter's hand toolbox used in similar form probably by every ship carpenter who ever worked in a wooden ship yard. The carpenters kept their complete collection of tools in their big chests stored in the shop. The tools expected to be used for the day were carried to the job in the hand box.

ONCE more we see the little schooner as she has now taken her form amid this lovely postcard setting at the upper bend of the Essex River. The spot is an historic one, having been a shipyard since the 1650s. Literally hundreds of wooden vessels have been built here, some of them almost as long as the little river was wide. On one occasion, nine boats at one time sat in this yard. Since 1813 it has been operated by members of the Story family.

The stanchions which Phil Terrio was working on are in, and another batten marks where the rail cap will be. The stem-head bears a chalked circle indicating the cut to be made for the bowsprit.

PLANKING the outside of a ship was really only half the battle. The inside had to be planked, too. Inside planking was known as "ceiling" and putting it in was quite as laborious, although perhaps not as painstaking as hanging the outside planking. You slipped and skidded as you worked and it always seemed as if your staging was too low (or too high). Furthermore, all those heavy planks had to be passed in over the outside of the ship. Here we see Steve Price, John Mulcahy, and Bob D'Entremont bending in a fresh piece of ceiling. In big ships it took special clamps called "ginneys" to do what John and Bob are here doing simply by leaning against the plank.

Ceiling contributed nearly as much to the strength of the hull as the outside planking did. In most cases the vessel was not ceiled all the way to the keelson or to the deck so that air could circulate freely in the space between the ceiling and the planking to inhibit rot.

THIS view shows the inside of a large dragger as the men are installing what is known as the "shelf." The shelf was just what the name implies: a laminated member which protruded from the inside upper edge of the hull and upon which the deck structure rested. It was installed against the upper two streaks of ceiling known as the "sheer clamps." It might be noted that this hull was not typical insofar as it has not yet been ceiled, for some unknown reason. We do see, however, that the bed for the heavy diesel engine is already installed.

WITH planking well along, the caulkers have begun their work filling the vessel's seams first with a "thread" or layer of cotton and then with two threads of oakum. For those who don't know, oakum is tarred hemp. On the left, Bill Boutchie is driving in the thread of cotton using the traditional caulker's mallet. To the right, Luther Burnham is using a big beetle to drive home a row of "rimming irons" (probably a corruption of "reaming irons") used to wedge open what is obviously a very tight seam. (Something must have happened with the beveller here.)

We know it must be cold since both men have their ear-laps down and Bill has his coat tucked into his overalls. For cold-weather wear, Luther was always partial to the old-time country-man's felt-topped boots, usually termed "felts."

THE planking of the little schooner of which we have been speaking was fastened with ship spikes rather than trunnels, and since the head of each spike is countersunk about 3/4″, it means that every spike hole must have a pine plug or "bung" inserted over it to finish off the job. Here is Pete Hubbard "bunging." First dipping the bung into white lead or thick paint, he inserts the bung making sure grain of bung corresponds to grain of plank. Fetching it a good fair blow with his mallet he drives it home. Then with a "slice" (a great big chisel) he trims it off flush with the surface of the plank.

WHAT good is a ship without a rudder? The rudder of a large vessel is a large and heavy thing, and a great deal of work is involved in making it. The photo clearly shows how the pieces are laid up, one atop the other, each a little thinner than the one below. After being fastened with long bolts or drift pins (and it's no picnic to run an auger down through all of that and not have it come out the side) the whole thing is trimmed and faired with an adze. Here again, demonstrating his ability to do just about any shipyard job, is Steve Price, making the rudder of a 110-foot dragger.

NO words of ours could add anything to this picture. Here he is, Steve Price, a grand old man of the shipyards, as he pauses in the trimming of the rudder for this engaging portrait.

THE ceiling and shelf are in, the stanchions in place, and the last major process of construction is underway—the framing of the deck. In this big schooner, we see the beams of the forward deck in place as the gang starts the raised deck aft. In fitting a big beam (roughly 8″ x 9″), one man works at each end. One of the laminated members of the shelf sticks up a bit and each beam is cut to lock over it. When fitted, a beam is held in place with bolts.

OBVIOUSLY any vessel going to sea with men aboard must carry a supply of fresh water. The use of cypress wood fresh water tanks has been traditional for years. This photo shows two things: how a cypress water tank looked and how the shears were used in putting it aboard. A cypress tank would outlast the life of a vessel and the water would always be sweet. The tanks must be put aboard before too much of the deck is complete.

Phil Tarr guides it in over the rail.

THERE are many things to notice in this deck view aft on the little schooner. First of all we see that the outer edge of the deck is already laid. These wide pieces are cut to drop down over the stanchions and are called "covering boards" (also sometimes known as "planksheer"). They are made watertight by driving pine wedges all around each stanchion. In the middle, with a chalk line for reference, is the central piece of decking called the "king plank." The rest of the deck is laid outward from these. Notice also the liberal use of "lodging knees" at the outer ends of the deck beams. Sharp eyes can further detect the use of "doublers" bolted to the beams at the ends of the deck openings.

THE work may often be hard and heavy, but it needn't be unpleasant or unduly taxing when there is an easy camaraderie and friendly rapport among those doing it, as for example in this picture. Here we see Leander Doucette and Dennis Martin, both, as we have mentioned, natives of Nova Scotia or "down home" (the terms were synonymous). They are spiking down the last section of king plank in the new deck and appear to be having a pleasant time. Of course we might add that anybody who worked with Leander Doucette had a good time, and that plenty of laughter was an integral part of life in the Essex shipyards.

IT would appear that the gang has gone to dinner, leaving quite a mess on the forward deck. It's a good view, though, of the laying of the deck and the beginnings of the various deck structures. Forward is a forecastle companionway; next we see a skylight, and in the foreground is the opening for the cabin trunk. The white pine decking seems to be about 2″ x 2″. Most decking was traditionally white pine, although fir was sometimes used.

The observer can count five hand toolboxes sitting around, albeit most of the tools seem to be lying on the deck.

ALL decks were fastened with spikes, countersunk, and bunged. Spikes were driven down flush and then, with the use of a "set," were punched home, snugging the deck down as they went. Edgar Boutchie, a set in his left hand and a five-pound maul in his right, is doing just this. The funny-looking wooden block with a handle on it is a "whammy" for driving in a last tight-fitting piece of decking. Instead of hitting the pine decking with the maul, you hit the whammy. Imagine the "hammer blossom" a five-pound maul would make on a piece of pine decking.

THE building of wooden vessels in the traditional way has always been a fascinating process to watch, more especially so in later years when, perhaps unconsciously, people realized that it was an industry which would not be around much longer.

In Essex the rather pastoral setting of our shipyards made it all very picturesque as well as interesting. This young man is but one of a great many who have made the building of a ship the subject of his painting.

A GREAT deal has been accomplished on the deck of the little schooner since the previous photograph. Looking aft we see that Marshall Cogswell, in straw boater, has the main cabin trunk well along while beside him Sammy Gray is puttying some of the deck seams. In felt hat, boss Jake Story seems to be delivering a last few words of particular instruction to Phil Terrio and Leander Doucette as they build the top of the transom stern.

THIS assembling of the traditional curved elliptical transom stern is one of the things that separated the men from the boys as far as ship carpenters were concerned. It was a fussy and painstaking job, no doubt about it, and it was a heavy and hard job too, as you can see from the size of the timber which went into it. If the oak timbers were this size for a 55-footer, imagine what they must have been for a big schooner. Tradition has it that the process was invented in Essex.

Master of the situation is Phil Terrio, here putting an auger into an electric boring machine.

THE stern on a dragger is totally different in configuration and construction from that of the schooner. In essence it is a roughly semi-circular structure built up from layers of heavy pine timbers sawed to the desired shape. Warren Sanborn, "Doc" Stanwood, Walter Cox, and foreman Willy Ross are fitting these timbers together. Stretching away behind the vessel is the little Essex River as it begins its winding path through the beautiful marshes to Ipswich Bay. Over 4,000 vessels "born" in Essex have taken this one-way voyage.

IF we stand on the ground, the dragger with the round stern we've been talking about looks like this. The jungle, or jumble if you will, of staging under and around the vessel is really impressive. Here in the James yard the vessels stand right next to the street.

JUST for fun let's go back a few years to the turn of the century and see a fishing schooner in somewhat the same stages of completion as the little schooner yacht and dragger of the previous pictures. In those days the appearance of a photographer was a real occasion, prompting the men to leave their work and pose for posterity. Note that the general appearance of things has shown little change over the years. Instead of the bright summer scenes we've looked at, however, the Essex River is here deep with ice.

It sometimes comes as a shock to those inquiring about life in the Essex yards to discover that shipyard work went on just the same all year around. The answer was that in those days a man was paid only for the time he worked—no work, no pay. (This, of course, was not unique to shipyards.) Thus, since most of our men needed also to eat in the winter, they worked in the winter, and, being outside as they were, it was rugged and disagreeable for much of the time.

THROUGHOUT our brief narrative we speak about the "gang." In contradistinction to the present-day connotation of the word, "gang" to us implied that group of men of a particular shipyard united in a sort of informal fraternal association having a common cause—the building of a good vessel and the making of a living while doing it. And there was a loyalty to one's yard that went with it; you were a "James man" for example, or a "Story man." Except for a few specialists, perhaps, there was little transferring from yard to yard, and such rivalry as may have existed was altogether friendly. Among the members of any one gang there was rarely any animosity. The rigors of this life of toil created a kind of brotherhood which tended to emphasize the lighter side of life wherever possible.

Since we seem to have no group pictures of the men of whom we speak in these pages, here is a priceless one from the days of yore. A. D. Story's gang in the days before the first World War included such men as Mose Bushey, Pete Hubbard, "Mac" MacIver, Allen Bruton, George Weston, R. Z. Davis, Willie Scroons, Jack Doyle, John Murphy, Stan Burnham, "Skeet" Doyle, Harry Swett, Owen Lantz, and Angus MacDonald. Behind them a big halibuter is all planked up.

A NICE view of the Lyman James yard with one of our growing draggers in the background. To the left is the mill with the band-saw out front and trunnel lathe and grindstone inside. Tool chests were also kept in there. In the foreground is a pile of oak plank and to the right is the old steam box where planks difficult to work were cooked. Steam was produced for this one in a big iron pot boiler bricked in underneath. In the pathway we see the crude two-wheeled wooden dolly-cart used to move heavier timbers about the yard. In earlier years a horse was used to drag the timbers.

The curious wooden device appearing in front of the mill at the left edge of the picture is the wheeled carriage used to support the ends of long planks or heavy timbers as they passed through the band saw.

JENNEY

PORTRAIT in an Essex shipyard—builder Lyman James and his foreman Will Ross. From the ground on which they stand to the yard with its ship standing across the river in the background, the area is steeped in the tradition of wooden shipbuilding for countless generations behind them.

WHETHER the deck view is of the little schooner or of a dragger such as we have here, it doesn't seem to make much difference. The clutter of odds and ends and assorted tools and equipment is about the same in either case. This photo rather interestingly shows the building of a "coaming" on a main fish hatch and illustrates the manner in which the corners were dove-tailed together. It's about all a man can do just to lift some of those big C-clamps seen lying there on deck.

ON a dragger a raised structure or deck called the "whaleback" is built over the forward end of the main deck. It becomes a place for lockers for the sheltering of miscellaneous fishing gear. It also keeps the main deck dryer in heavy weather. Dennis Martin is here starting to trim the insides of the main frames and forward fillers in preparation for the work. Behind him lying on deck are two round-faced adzes, and near his left toe is a ship carpenter's big "slice" and a caulker's long-handled "horsing" iron. The steeple of the Universalist Church peeks around the corner.

BY this stage of the game, the "inboard joiners," artistocrats among ship carpenters, have begun their work. These men built the living quarters and accommodations for crew and the structures on deck which pertained to them. For example, we see that Nelson Lane is here finishing up the forecastle companionway. He will afterwards go below to build the forecastle itself.

IT was usually expected that, when a vessel reached the degree of completion illustrated in this picture, the owner would begin to ask when she was going to launch. However, some of the fussiest and most time-consuming of the jobs still remain to be done.

Way forward the gang is getting the deck beams onto the whaleback; 'midships, Mike Doyle is trying to sand up the sides of the forward main hatch but yard owner Lyman James (in white shirt and galluses) is talking a blue streak; aft, Nelson Lane, having finished the forward companionway, is getting out the 3″ white pine sides for the main cabin trunk. Warren Sanborn is just coming out of the hold with a template for a floor timber. He's probably been down there working on the bed for the main engine.

Calmly surveying this peaceful summer scene are the two steeples, as they have been doing for generations. The Congregational steeple to the right contains a Paul Revere bell whose tones once signalled the beginning and end of the working day in the Essex yards. In 1800 that day lasted from sun-up to sun-down.

FOR a closer look at the makings of the whaleback we see this shot of the beams as they are fitted and placed. The structure is quite as heavy as that of the main deck itself. Origin of the term "whaleback" is obscure. They were sometimes referred to as "forecastle heads."

ONE glance at the caulker with the derby hat in the right foreground tells us we're back in the "good old days." In fact, we're back in A. D. Story's yard in the summer of 1912. We're looking at one of the few photographs ever made of work in progress in an Essex yard of that period. The deck so interestingly portrayed here is that of the *Knickerbocker,* a typical big Gloucester schooner. Aft we see the decking being laid on the main cabin trunk; in fact we see a caulker up there already caulking some of it. The big black spots on the side of the cabin trunk are where the knots in the white pine have been shellacked before priming. Lying amidships on the port side are bundles of the heavy galvanized iron chain plates with their lignum vitae dead-eyes. The *Knickerbocker* ended her career on the Pacific Coast.

Across the way is the James shipyard and further beyond is the village of South Essex.

IN the previous picture of the *Knickerbocker* we noted the construction of the main cabin trunk. In all probability the man who built it was the man we see here in this photograph, taken twenty-nine years later, in the process of doing the same thing. He's Ed Perkins, by common consent one of the finest inboard joiners ever to work in an Essex yard. Ed's work was the standard by which all other work was judged. Here, in his early eighties, he is finishing off a cabin trunk of a big dragger and preparing to build a good-sized pilot house on top of it. The wooden fore-plane on the deck in front of him and the short jointer he's using he made himself.

Occasionally, when not working in the shipyard, Ed built houses and in his spare time made furniture of exquisite workmanship. When asked, once, by a shipyard colleague how well a particular job needed to be done, Ed replied, "You do the very best you can and when you get done it won't be any better than it ought to be!"

JOHN MULCAHY walks aft past Leander Doucette, who with mallet and gouge is starting to cut the hole for the mainmast. (He has first outlined the cut with a ring of auger holes.) Pointing at us from forward is Marshall Cogswell's straw hat as he stoops to apply some sort of finishing touch to the forward companionway. At the left, the end of somebody's hand box shows a flair for the Gothic. Auto buffs will spot a 1934 Packard at the right edge of the picture.

AGAIN we see John Mulcahy, who holds one of the small cast iron hawse pipes that he's about to fit through the "knight heads." Technically, I suppose, we shouldn't really call them knight heads; actually they're fillers. Strictly speaking a knight head is only the piece against the stem itself. The fillers, to which the term knight head came to be loosely applied, filled in the rest of the space back to the first stanchion. Anyway it's through these fillers that the hawse holes for the hawse pipes were cut, and it's through the hawse pipes that the anchor cable ran out. To John's right stretches the newly installed bowsprit.

IT'S Sunday and the little schooner rests quietly in the yard while she waits for the return of the gang on Monday morning. The bowsprit and hawse pipes are in and most of the rail cap is on. A good portion of the topsides are smoothed and primed.

The chips on the ground must go down a fathom or more since ships have been built on this spot for nearly 300 years.

INEVITABLY as soon as a deck is laid, a mess appears on it. The owner himself wields a broom up forward as he sweeps an area clear so that it can be caulked. On either side we see rail caps being applied, while amidships appears the main cabin trunk. It's hard to say what that bushel basket is doing there.

SPEAKING of rail caps reminds us that it's no mean job to put these on. They must be "molded" (shaped, that is) from wide clear pieces of heavy oak plank and then mortised down over the tops of each stanchion. All joints must needs be a long scarph. When on, all must be trimmed and smoothed and edges nicely rounded.

Here Walter Cox is getting out what appears to be the last piece of rail cap to fit the middle of the dragger's stern. He's using the starboard quarter bitt as a prop for one end of the piece. We see some more "rimming irons" left sticking into a deck seam. Nearby lies a caulker's curved iron for caulking a seam which runs up against a vertical surface.

BY now some of the mess on deck is gone and Marshall Cogswell has the canvas on the trunk-top. On smaller boats and often on yachts, trunk-tops were covered with canvas instead of decking. The transom is almost finished too. In the foreground a skylight shows in fine detail. There's that bushel basket again. Oh yes, and we see some chain plates lying under the starboard rail. Decking in the foreground is now all caulked and the seams puttied.

LUTHER BURNHAM was an exception among caulkers. He caulked decks working on his knees. Most caulkers sat on their little sliding combination seat and toolbox or else used a small folding stool that ran on wheels. Whether kneeling or wheeling, the tools were the same and so were the methods. A caulker's black mesquite mallet was practically an extension of his own right arm. Here we have a fine detail of a caulking mallet. We also see a caulker's reefing hook, chisel, and bevelled iron.

Perhaps more than any other sound, the ringing of the caulker's mallet striking the caulking iron told the whole town a ship was being built.

WHEN it came to cutting in portlights, nobody could beat Stanwood Burnham, shown here applying a liberal layer of white lead before setting in the heavy bronze outer rim. Lots of people never knew his real name was Stanwood. To everyone he was simply "Needles." In an age not as hirsute as our own, Needles was distinguished as the man who shaved once a week—every Saturday. One might also observe that maybe the time has come for Needles to throw away that old coat he's wearing in favor of one not quite as ancient.

Just beyond Needles' cap is one of those sliding seats used by a caulker.

IN 1941, the year of our picture, traffic along Main Street in Essex was light enough that one could easily step across and compose the fine broadside view of our dragger nearing completion in the James shipyard. The absence of all but one lone figure under the bilge amidships tells us it must be the noon hour. We do get, however, a good look at the size and shape of one of these boats. We also see the rudder, finished and ready to hang, and the framing of the pilot house.

A SIMILAR dragger in a similar state of completion (and in the same year) stands across the way in Story's. The pilot house we see here is the one which Ed Perkins in an earlier picture was getting ready to start. That's Ed up there at the right. Just abaft the break in the deck are the two hand-operated bilge pumps, a type carried by all fishermen from time immemorial. We note further that all hatch covers have been completed and are carefully leaned against the hatch coaming or lying face up. Once made, a hatch cover is never placed upside down—a sure sign of hard luck for the vessel.

FINISHING touches" is one of the more hackneyed phrases used to describe pictures illustrating the completion of almost anything, be it building, bridge, or boat. However, it is about the only one which can be applied to this one as the gang swarm all over the topsides, and Sammy Gray, alone under the little vessel, puts on the last few brushfuls of copper paint. Launching day cannot be far away.

EVEN a dragger needs a set of good stout masts, and the man to make them is Arthur Gates. Starting with a carefully selected white pine tree, Arthur must first scrutinize it from all sides, imagining as he does so the finished mast contained within. Taking into account any degree of natural sweep and the minor surface imperfections which any tree has, he lays out an initial centerline. From this he can line out the taper from top to bottom. Hewing the tree to these lines gives him two flat sides. Repeating the process of selection he can now again line out the mast in a plane at 90° to the original, ending up with a square section and with many of the imperfections eliminated. In the picture above this is what he has done with the stick in the foreground. The next step is to hew the square to an octagonal section and then with drawknife and plane, the round finished mast emerges.

FOR big heavy cuts, Arthur would be holding the end of the axe-handle and swinging nearly over his head. Here he's trimming a little more closely and has choked up about halfway on the handle. Depending on whether the owner hews right or left handed, a broad-axe handle is offset one way or the other.

This particular pine tree has come from New Gloucester, Maine. Arthur Gates made his masts in whichever of the yards happened to be his customer. Behind him here we see the top end of a new keel and the mill and shop of the Dana Story yard.

A "SHUTTER" is that last piece of planking to be put in which completes the planked surface. With a grin of satisfaction, Steve Price and Leander Doucette have just spiked off the little shutter under the rail cap, completing the bulwarks of the *Blackfish*. (The little schooner has a name now.)

AT last! She's done—well, almost done. There's still a little deck work left and she must be rigged, but the hull of the *Blackfish* is finished and stands ready to be launched. Underneath, Phil Terrio and Steve Price busy themselves with ways and cradle while Leander Doucette, sitting on the bow in a remarkable representation of Rodin's *The Thinker,* contemplates where and how to make fast the snubbing line attached to the "drag." Bob D'Entremont is climbing the ladder on one last errand.

OBVIOUSLY the launching of the vessel is the most spectacular episode of the building process. Those few exciting moments are preceded by days of careful preparation and labor, since a great deal is at stake.

There are two ways of doing it. First of all is the traditional method of building a cradle (perhaps "sled" would be a better word) under the vessel and letting her slide down a greased track or "ways." The second method, and the one perhaps more commonly used through the years in Essex, is the one illustrated in the next pictures. In this procedure, the vessel slides on her keel across a succession of greased slabs and with a single greased sliding way built up under one bilge.

Here, Jonathan Story and his men are lining up the row of slabs across which the keel will slide. One of the more irksome of these tasks is the necessity for keeping this mess of timber from floating up at high tide. Long and heavy stakes driven into the mud usually suffice.

PHYLLIS S.

THIS view shows the single way upon which the bilge will slide and how it has been built up and braced. The weight of the vessel is brought to bear by leaning her over with heavy screw jacks. We called this method a "side launching," and, as you see, it does *not* mean that the vessel took the water sideways. It was done this way for several reasons, not the least of which was the fact that it was considerably less work and consequently cheaper; also the vessel took a lot less lumber and trash into the water with her. A third point which was of importance in some situations was that in going into the water on her beam ends, the vessel drew considerably less water.

Note the heavy bracing on the rudder to keep it from accidentally being twisted as the vessel goes in.

RONALD & MARY JANE
NEW YORK

TO accomplish a side launching, the blocks upon which the vessel has been sitting during her construction must be split away in order to bring the keel to bear upon those greased slabs we saw. The exciting part comes in not knowing how many blocks must be split out to release the boat. The men start splitting at the after end and work forward. As they approach the bow the tension mounts as often the creeping of the vessel becomes visible. She wants to go. Then, with two or three blocks remaining, a sudden crack and a groan and the vessel is on her way.

Occasionally all the blocks must be split before she moves. Such is the case in this picture as Clarence Mulcahy and Sammy Gray put the splitting wedge to the last remaining block. As the ship begins to move they will leap clear.

It's all very dramatic.

BECAUSE she was small with high bilges, it was easier to launch *Blackfish* in a cradle. Interestingly, the same method was also used on the largest and heaviest boats we built, the details and methods of procedure being exactly the same.

The runners of the cradle or sled are known as the "sliding ways" and the tracks down which the cradle slides are known as the "ground ways." Before release, both sets of ways are clamped and/or bolted together at the upper ends. To release the boat it is only necessary to saw off the sliding ways just behind the place where they are fastened. A pair of men with a "cross-cut saw" on each side do the sawing, making sure as they go that both way ends will be severed at the same moment.

This is what Pete Cogswell and Phil Terrio are doing in this picture. Unfortunately we can't see who's sawing on the other side. The sponsor with the bottle of champagne awaits the crucial moment.

THIS would seem to be the "crucial moment" for this boat. The ways are severed, the boat begins to move and "crack" goes the bottle. It looks as though the sponsor and Clarence Mulcahy will get a face-full of champagne.

"THERE she goes!" The moment of truth has arrived. The period of gestation is over and the ship is born to become a creature of the sea.

Amid waving of caps and salutations of joy, the schooner yacht *Skilligolee* takes the waters of Essex River. (We used the *Skilligolee* here because it's such a good shot and works in more easily with the following picture.)

EVEN though he is back to us, one can almost sense the feeling of relief experienced by builder Jake Story as, with hat in hand, he gazes at his latest creation, now safely water-borne. From Essex she will be taken in tow for Gloucester, where machinery and rigging will be fitted. Notice all the lumber that a cradle launching carries into the water with the vessel.

HAVING watched so much of her construction and seen the building of her ways, it seems only fair that we should watch *Ronald and Mary Jane* take her initial plunge using the traditional "side launching." Some idea of the enthusiasm for these events can be gained by the size of the crowd pressing every vantage point for a better look.

IF you want to see what an old time hair-raiser of a side launching looked like, here's the *Emerald* breeching the tide from Story's yard in August of 1924. If you were on deck you hung on for dear life. The smoke at the bow is from the hot launching grease, which occasionally caught fire from the frictional forces.

PERHAPS no more fitting conclusion to our picture story could be found than this splendid shot of *Blackfish* now afloat on the Essex River, taken through the frames of the vessel following behind her. An end and a beginning. A cycle repeated in Essex nearly 4,000 times over a span of 300 years.

THE great majority of the men who labored with heart and soul and aching backs to build the Essex vessels were never able to see their ships in the true environment for which they were intended, much less be aboard them in such a circumstance. There must have been many, however, who, through the long hours of hewing, adzing, planing, fitting, boring, caulking, painting, or whatever, dreamed of the day when the vessel would be out there in the midst of great seas and the worst of the elements. Perhaps as they did so they put just a little more care and pains into the job, remembering the lives dependent upon the skill of their hands.

Here we see what it all came to. The last workman has picked up his tools and gone ashore and the mariners have taken over where the shipwrights and riggers left off. This is the *Gertrude L. Thebaud* on the port tack in a good stiff breeze off the back shore of Cape Ann.